Do you face a tough problem? Are you in a very rough spot? Do not become sad or nervous. Let Blue, the dog, help you out. Let Blue come to the rescue!

These women need to cross the street. Cars, vans, and trucks rush by. It is a very busy place. Both women have become nervous. Do not be nervous, women. Blue is due here soon!

Here is Blue to the rescue! He'll come through for sure. Blue leads both women through the busy street. Now the women are joyous, not nervous. Blue is a true pal.

The man has lost his blue hat. He cannot pursue it in the wind. That is a tough problem for him. The blue hat is of great value to him. He is in a rough spot for sure!

Look! It is Blue to the rescue again! Blue can pursue the blue hat. Blue comes through for the man. Now he is a joyous man. Blue is a true pal for sure!

Sue wants to make a birdhouse. First she needs to measure the wood. How can Sue measure it? Here comes Blue to the rescue! Blue brings what Sue needs.

Next, Blue brings glue to Sue. Then Blue brings the paint. Blue is a very busy dog! The rough and tough job is now through. Sue and the bird are joyous. And it is all due to Blue. How marvelous!

Blue gets all the praise he is due. He is well known all over town. Blue is a true pal to all. You cannot argue with that. Do you face a rough or tough problem? Be joyous, not nervous. Let Blue come to the rescue!